SpringerBriefs in Molecular Science

History of Chemistry

Series Editor

Seth C. Rasmussen, Fargo, USA

For further volumes:
http://www.springer.com/series/10127

Jacobus Henricus Van't Hoff (1852–1911, Nobel 1901)
Grandfather of Polymer Science
Edgar Fahs Smith Collection, University of Pennsylvania Libraries

Gary Patterson

A Prehistory of
Polymer Science

 Springer

Gary Patterson
Department of Chemistry
Carnegie Mellon University
4400 Fifth Avenue
Pittsburgh, PA 15213
USA
e-mail: gp9a@andrew.cmu.edu

ISSN 2191-5407 e-ISSN 2191-5415
ISBN 978-3-642-21636-7 e-ISBN 978-3-642-21637-4
DOI 10.1007/978-3-642-21637-4
Springer Heidelberg Dordrecht London New York

Library of Congress Control Number: 2011940054

Springer is part of Springer Science+Business Media (www.springer.com)

Preface

The present volume examines the time period before there was a coherent scientific community devoted to the study of macromolecules. It starts with a series of studies of particular polymeric materials that were important in this pre-paradigm period. The history of natural rubber is followed from the time of the great French explorers (1735) to the formulation of the first successful theory of rubber elasticity (1935). The history of polystyrene is presented from the discovery of styrene in the late eighteenth century to 1935. The first commercially successful polymeric material synthesized completely from inexpensive small molecules, Bakelite, provides a fascinating story of both academic and industrial chemistry. The story of the polysaccharides and Sir Norman Haworth (Nobel 1937) completes the studies of materia polymerica. The crowning event in the prehistory of polymer science is the Faraday Society Discussion of 1935 on Polymerization. A history of the Faraday Discussions that led up to this event is presented. The chronicle of the Faraday Society includes Discussion Meetings that went from glory to glory until the Society was absorbed by the Royal Society of Chemistry. The book concludes with an essay on the prehistory of polymer science. The factors that influenced this history form a fascinating study of the formation of a now thriving scientific research community.

Gary Patterson

Acknowledgments

My interest in polymer science was kindled by Paul J. Flory as an undergraduate and fanned into flame while earning my PhD in his laboratory at Stanford. He set a high standard both for science and for writing.

My serious interest in the history of polymer science was stimulated by Herbert Morawetz and his ground breaking monograph: Polymers—The Origins and Growth of a Science. He has been a constant source of encouragement.

I was welcomed into the community of chemical historians quite late in my scientific career by John Hedley Brooke of Oxford and Seymour Mauskopf of Duke. They also helped me to obtain a Fellowship at the Chemical Heritage Foundation in order to learn my craft. Arnold Thackray, Mary Ellen Bowden and the staff of the CHF instructed me as a novice historian. I have continued to work at and with the CHF ever since. I am currently the Chief Bibliophile of the Bolton Society.

The HIST division of the American Chemical Society warmly received me into their company. Many talks and symposia later, Seth Rasmussen joined Elizabeth Harkins of the Springer Publishing Company in proposing a series of books on Chemical history. Authors from HIST were invited to submit books to this project. The present book is the first fruit of this vision. Many of the sections are based on actual talks given at ACS Meetings.

The basis of a life as a scientist and historian is a warm and supportive home life. Susan has joined me in my strange life for more than 44 years. It would have been impossible without her.

Contents

Chapter 1
Introduction

Polymer science is now a thriving multidisciplinary scientific community. It is composed of scientists, engineers, technologists and industrialists from a very wide range of academic communities: mathematics, physics, chemistry, biology, chemical engineering, materials science and engineering, mechanical engineering, plastics engineering, dentistry, textile engineering, and many more. What unites this disparate group is a belief that the material world contains substances that can best be understood in terms of extended macromolecular structures: polymers. While the word **polymer** simply means "many repeat units", and while many actual substances that are not molecular in nature have been referred to as polymers, the present treatise will focus on the growth in understanding of substances that are genuinely macromolecular.

Since the Earth on which we live is filled with macromolecules, it might be supposed that an understanding of these substances would date from antiquity. However, most natural materials are mixtures, and detailed studies of these substances needed to wait until Chemistry had progressed to the point where reproducible analyses could be carried out. I have chosen to start the story in the eighteenth century, when explorers brought back to Europe many fascinating substances and placed them in the hands of the best scientists of their day. These materials were also eagerly embraced by technologists who wanted to produce items that would benefit humankind and enrich themselves. The free flow of materials and information that characterized the French community produced much great science and not a little benefit to humankind. The first materia polymerica chosen is caoutchouc, now routinely known as natural rubber.

Tree saps have played an important role in human culture. Styrene was discovered by heating Storax resin, and polystyrene was discovered by letting styrene sit undisturbed for a few months. Polystyrene has played a central role in the intellectual development of polymer science, but it did not become a commercial material until after the time period of this treatise.

The first commercially successful polymeric material synthesized entirely from inexpensive small molecules was Bakelite. The story of the discovery, development and understanding of Bakelite is a fascinating combination of pure

G. Patterson, *A Prehistory of Polymer Science*, SpringerBriefs in History of Chemistry,
DOI: 10.1007/978-3-642-21637-4_1, © The Author(s) 2012

chemistry, applied chemistry and shrewd technology. Bakelite is a materia polymerica for the ages.

Another ubiquitous earthly material is the class of carbohydrates. They have been important to human civilization from the beginning. The understanding of the nature and structure of pure polysaccharides came relatively late in the prehistory of polymer science, but the first Nobel Prize associated with macromolecules was awarded to Sir Walter Norman Haworth in 1937.

While individuals labor as scientists, they do so in the context of a community that considers the reported "facts", evaluates the proposed theories and allocates honors and resources. A fully visible community of scientists devoted to the macromolecular paradigm did not exist until the 1930s. In order to help understand the emergence of this community, a brief history of the Faraday Society and its Discussions will be given. A celebrated Discussion on polymerization in 1935 produced the kind of consensus that results in a large group of committed workers. The group photograph of this meeting includes most of the people that went on to define the initial paradigm period of polymer science.

A concluding essay discusses the prehistory and evaluates the various forces that influenced the history of polymer science. Actual human culture is driven by multiple factors and is both more complicated and more interesting than ideological stances. Polymer science provides a rich milieu in which to explore the development of a scientific community.

Chapter 2
Materia Polymerica

2.1 Natural Rubber (Presented to HIST, March 2011)

One of the great French explorers of the eighteenth century was Charles Marie de la Condamine (1701–1774). During his explorations in South America, he encountered the substance now known as natural rubber while in Ecuador [1], and sent back samples to the Academie Royale des Sciences in 1736. De la Condamine presented a paper by his collaborator Francois Fresneau (1703–1770) before the Academie in 1751, describing some of the properties of the substance dubbed "caoutchouc" [2]. Another Frenchman from the Academie Royale des Sciences who studied caoutchouc was Pierre Joseph Macquer (1718–1784). He was the foremost French authority on Chemistry after Lemery and before Lavoisier [3]. Macquer's most famous work was his "Dictionnaire de Chymie," which was initially published in 1766 and was revised and reissued in 1778. In 1763 Macquer published a paper in the Histoire de l'Academie Royales des Science on his own studies of caoutchouc [4]. One of the goals of this early scientific work was to purify the substance and to discover a suitable solvent, so that it could be easily processed. Macquer, like Fresneau, found that turpentine was a good solvent and that cast films retained their elasticity. Macquer was also able to dissolve caoutchouc in carefully distilled and dried (rectified) ether [5]. Since ethyl ether does not occur in nature, it is notable that Macquer had a sufficient supply to use it as a "solvent."

Soon after caoutchouc was introduced in England, Joseph Priestley (1733–1804) in 1770 discovered its use in removing pencil marks from paper [1], which led to its nickname: "rubber." People have been finding new uses for caoutchouc ever since.

Caoutchouc began to appear in large compilations of chemical knowledge such as J. Murray's "A System of Chemistry" [6]. It was noted that the tree sap from the *Hevea guianensis* (a type of Euphorbia) could be separated into a firm elastic coagulum and a watery liquid. The specific article on caoutchouc contains a clear declaration: "The most remarkable physical property of which this substance is

G. Patterson, *A Prehistory of Polymer Science*, SpringerBriefs in History of Chemistry, 3
DOI: 10.1007/978-3-642-21637-4_2, © The Author(s) 2012

possessed, and which eminently distinguishes it, is its high elasticity." [6]. The article in Murray also called attention to one of the most remarkable studies of caoutchouc carried out in 1803 by the Blind Philosopher of Kendall (John Gough (1757–1825)) and published in the Memoirs of the Manchester Literary and Philosophical Society. When a strip of caoutchouc was rapidly stretched, its temperature increased. Gough was well known for his ingenuity and used his lips to detect the rise in temperature. An even more remarkable experiment was described where a strip of rubber under tension from a suspended weight was heated. The strip was observed to shorten under heating! Gough was a great teacher and influenced another Manchester notable, John Dalton! Another chemical test was dry distillation, one of the classic protocols of alchemy. Natural rubber yielded an empyreumatic oil under these conditions.

Caoutchouc soon became an item of commerce. The firm of Thomas Hancock (1786–1865) developed a line of products based on both the elasticity and the water repellency of caoutchouc [7]. In order to learn more about the substance that was the basis of his wealth, in 1824 Hancock gave some caoutchouc to his friend Michael Faraday (1791–1867) at the Royal Institution. Faraday was the foremost analytical chemist of this era. As was his custom, Faraday repeated the known experiments on Hevea tree sap, verified the ones that were true and became proficient in the purification and processing of caoutchouc [8]. A copy of the title page of Faraday's classic "Experimental Researches in Chemistry and Physics" is shown as Fig. 2.1.

When fully purified and prepared as a solid block, caoutchouc is transparent. The good optical quality of pure caoutchouc led Faraday to carry out optical experiments. Relaxed and annealed caoutchouc is isotropic, both physically and optically. When stretched, it becomes birefringent. The production of optical anisotropy fascinated Faraday. It was also observed by Sir David Brewster (1781–1868), the noted optician.

Another remarkable property of the block of purified caoutchouc occurred when it was stretched even farther. The block became hard and very tough. This form could be heated and the block returned to its initial shape and elasticity. It is now known that the phenomenon being observed is crystallization, but at the time it was noted and puzzled about. Rubber has been a source of amazement and amusement ever since its discovery.

Faraday also heated the rubber in an anaerobic distillation apparatus, as described in Murray's account. A group of volatile substances were isolated. He was able to show that caoutchouc is a pure hydrocarbon, with no traces of oxygen or nitrogen. Elemental analysis yielded a C/H ratio of 8/7 in early nineteenth century practice. (Carbon was then treated as $M = 6$, rather than 12, with H as 1). Modern values are represented by the empirical formula C_5H_8.

Another leading scientist that took an interest in caoutchouc was Andrew Ure (1778–1857). He published the greatly influential work "A Dictionary of Arts, Manufactures and Mines" in 1837 [9]. The article on caoutchouc ran from pages 357 to 366. In addition to the usual information about separation of the rubber

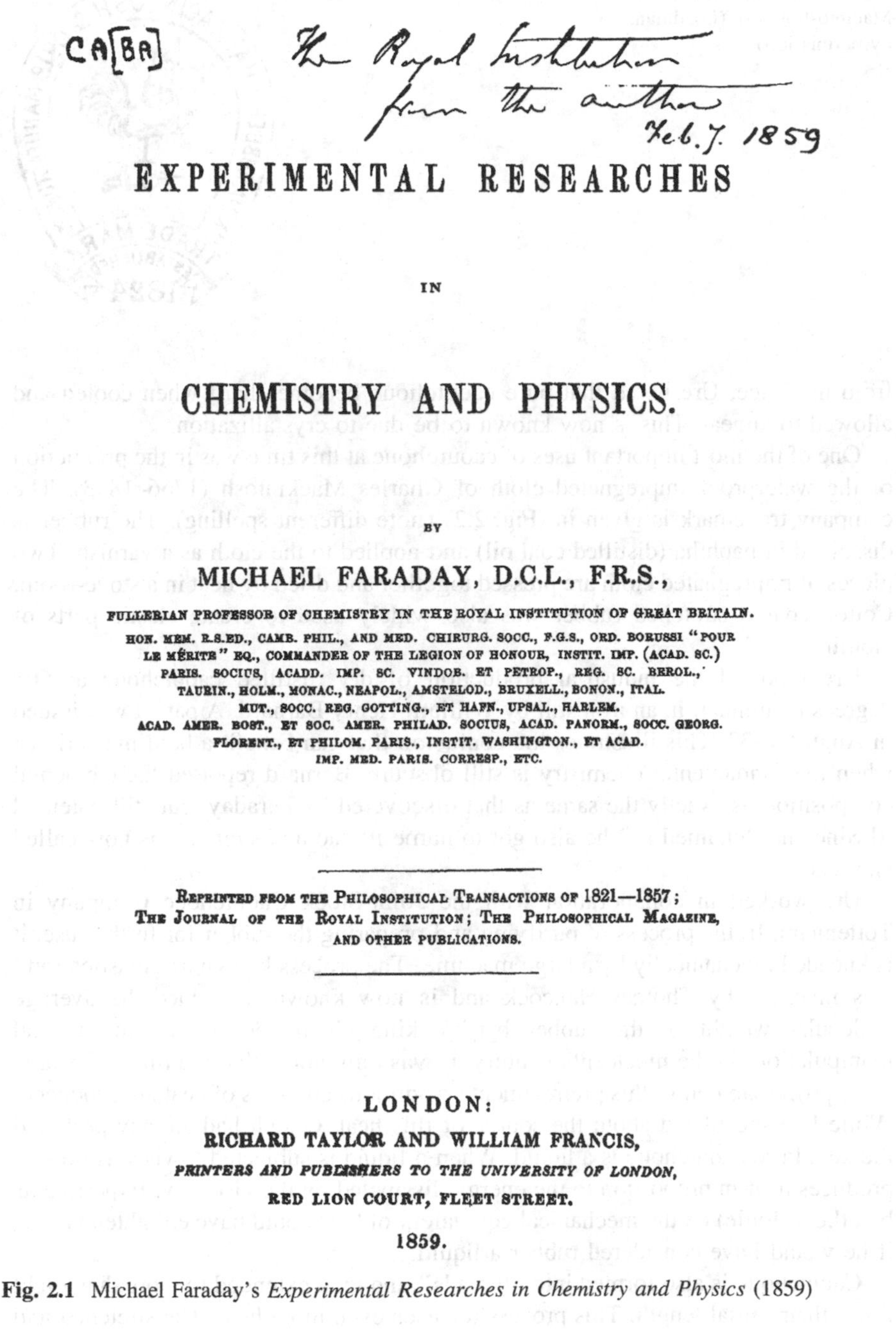

CA[BA]

The Royal Institution
from the author
Feb. 7. 1859

EXPERIMENTAL RESEARCHES

IN

CHEMISTRY AND PHYSICS.

BY

MICHAEL FARADAY, D.C.L., F.R.S.,

FULLERIAN PROFESSOR OF CHEMISTRY IN THE ROYAL INSTITUTION OF GREAT BRITAIN.

HON. MEM. R.S.ED., CAMB. PHIL., AND MED. CHIRURG. SOCC., F.G.S., ORD. BORUSSI "POUR
LE MÉRITE" EQ., COMMANDER OF THE LEGION OF HONOUR, INSTIT. IMP. (ACAD. SC.)
PARIS. SOCIUS, ACADD. IMP. SC. VINDOB. ET PETROP., REG. SC. BEROL.,
TAURIN., HOLM., MONAC., NEAPOL., AMSTELOD., BRUXELL., BONON., ITAL.
MUT., SOCC. REG. GOTTING., ET HAFN., UPSAL., HARLEM.
ACAD. AMER. BOST., ET SOC. AMER. PHILAD. SOCIUS, ACAD. PANORM., SOCC. GEORG.
FLORENT., ET PHILOM. PARIS., INSTIT. WASHINGTON., ET ACAD.
IMP. MED. PARIS. CORRESP., ETC.

REPRINTED FROM THE PHILOSOPHICAL TRANSACTIONS OF 1821—1857;
THE JOURNAL OF THE ROYAL INSTITUTION; THE PHILOSOPHICAL MAGAZINE,
AND OTHER PUBLICATIONS.

LONDON:
RICHARD TAYLOR AND WILLIAM FRANCIS,
PRINTERS AND PUBLISHERS TO THE UNIVERSITY OF LONDON,
RED LION COURT, FLEET STREET.

1859.

Fig. 2.1 Michael Faraday's *Experimental Researches in Chemistry and Physics* (1859)

Fig. 2.2 Trademark of Chas.
Macintosh & Co. (Loadman,
by permission)

from the juice, Ure noted that pure caoutchouc becomes hard when cooled and allowed to anneal. This is now known to be due to crystallization.

One of the most important uses of caoutchouc at this time was in the production of the waterproof impregnated cloth of Charles Mackintosh (1766–1843). The company trademark is given in (Fig. 2.2) (note different spelling). The rubber is dissolved in naphtha (distilled coal oil) and applied to the cloth as a varnish. Two pieces of impregnated cloth are pressed together and dried by heat in a stove-room. Cotton coated stretched rubber was also widely used to create elastic parts of clothing.

Ure reported the industrial production of dry distilled caoutchouc at 600 degrees Fahrenheit in an iron still by William Henry Barnard. A patent was issued in August 1833. This illustrates the confusion that can benefit a bold industrialist when the fundamental chemistry is still obscure. Barnard reported the elemental composition as exactly the same as that discovered by Faraday, but still patented it! Since he "claimed it," he also got to name it: caoutchoucine! It is now called isoprene.

Ure worked in collaboration with the Joint-Stock Caoutchouc Company in Tottenham. In the process of purifying and preparing the rubber for further use, it is kneaded mechanically by a large machine. The process known as "mastication" was invented by Thomas Hancock and is now known to reduce the average molecular weight of the rubber by breaking chains during the mechanical manipulation. In the nineteenth century, it was only noted that the rubber became more processable after this pretreatment. Prodigious amounts of heat are produced. While Ure speculated about the source of this heat, Gough had already provided the key. Pure caoutchouc is a liquid. When a liquid is subjected to viscous flow, it produces heat in proportion to the energy dissipated by the viscosity. Experiments by others (Joule) on the mechanical equivalent of heat could have enlightened Ure, if he would have considered rubber a liquid.

Caoutchouc is also formed into "threads" and then extended to more than eight times their initial length. This process releases even more heat. The stretched and recooled threads are then quite inextensible. The process is strain induced crystallization. The threads can be heated and allowed to retract and anneal, after which they regain their full elasticity. The specific gravity of the inelasticated rubber is 0.948732, while the reconstituted rubber is 0.925939. This is consistent with the higher density of the crystalline state of natural rubber.

The attempt to produce a true solid from pure caoutchouc was achieved by Nathaniel Hayward (1808–1865) of the Eagle Rubber Company and by Charles Goodyear (1800–1860) of New York. The process involved mixing sulfur and other substances with the rubber and applying heat. Hawthorne obtained a US Patent (#1090) in 1838 and Goodyear purchased all rights under the patent [1]. In 1842 Goodyear sent samples of heat and sulfur treated India rubber to England and distributed them to Thomas Hancock, now also associated with Charles Macintosh & Co., and to the Joint-Stock Caoutchouc Company. The British manufacturers wasted no time in "improving" the process and obtaining their own patents. The actual chemical process is now known to be due to crosslinking by sulfur bridges between the caoutchouc chains, but in 1839, when it was discovered, it just produced money! Hancock obtained a British patent in November, 1843, soon after the death of his close friend Charles Mackintosh. Hancock spent the next six months, as required by English patent law, perfecting the manufacturing process. The current name for the process, vulcanizing, was coined by William Brockedon, another director of Charles Macintosh & Co. Hancock was the foremost rubber technologist in England and had been manufacturing rubber products since 1820. He proceeded, like Faraday, to test all the variables of production and to optimize them. He produced vulcanized rubbers all the way from lightly crosslinked films to heavily crosslinked Ebonite. Ebonite is still used today to make "hard rubber" objects such as bowling balls, billiard balls and golf balls (Fig. 2.3).

The compositional details of caoutchouc, gutta-percha (an isomer of caoutchouc) and caoutchoucine were finally settled by Charles Greville Williams (1829–1910) [10]. In 1860 he carried out a substantially more careful dry distillation of caoutchouc in an iron alembic and isolated two volatile fractions. The lower boiling one (37–38 °C) had a (modern) molecular formula of C_5H_8, not just an empirical formula. He called it isoprene, to distinguish it from the crude fractions obtained by previous workers. A dimer was also isolated and characterized. All these materials had the same pure hydrocarbon empirical formula.

If the chemist could deconstruct natural rubber, could he also construct it? G. Bouchardat claimed to synthesize rubber from isoprene using HCl as a catalyst in 1879 [11]. Sir William Augustus Tilden (1842–1926) brought an even higher level of both synthetic ability and theoretical intuition to the problem. He synthesized rubber from isoprene and inferred the correct structural formula for isoprene in 1884 [12].

Organic chemistry was making great strides in the end of the nineteenth century and in 1897 W. Euler was able to prove the structural formula for isoprene

Fig. 2.3 Thomas Hancock in
1841, painted by T. Overton
(Loadman, by permission)

completely from synthesis [12]. This was the stated goal of Marcellin Berthelot
(1827–1907) in his groundbreaking book "Organic Chemistry Founded on Synthesis" [13]. The concept that was still unformed was the conformation of the
polymer: polyisoprene.

The early experiments of Gough were refined and completed by James Prescott
Joule (1818–1889) in 1858 [14]. Joule was another of the Manchester masters of
experimental chemistry. William Thomson (Lord Kelvin) had formulated the
fundamental principles of solid elasticity. Isotropic solids were characterized by
two elastic moduli: the modulus of compression, K, and the shear modulus, G.
Perfectly elastic distortions of the body stored the energy in the elastic potential
energy of the body. However, extensional deformations led to an increase in the
volume of the body; this required work against the internal pressures and led to a
decrease in temperature. Joule verified this effect for many typical solids. However, when natural rubber was subjected to uniaxial extension, the temperature was
observed to rise. The volume was almost unchanged during the deformation, so
that the normal cooling effect was minimized, but what was giving rise to the heat?

It was known that caoutchouc and gutta-percha (another purified tree sap) had
the same empirical formula. When gutta-percha was subjected to the same force
applied to the caoutchouc, it did not deform as much, and it cooled, as expected!
What was so different about these two otherwise chemically identical substances?
Crystallinity! The crystal deformed as an elastic body and the sample returned to
the same temperature upon release of the force. Joule subjected caoutchouc to a
cooling-bath at $0°$ Fahrenheit for a few days. The sample then became rigid and
showed the same elasticity as the gutta-percha. Joule worked like Faraday: he tried
everything!

Joule also carried out a systematic stretching of a piece of rubber with
increasing weights. He measured the initial length, the temperature change, and the

final length. He noticed that the sample deformed partly irreversibly. While this viscous elongation did complicate the analysis, it did not stop Joule from measuring the effect of temperature on the sample at fixed force. As the temperature increased, the sample shortened initially, and then extended viscously over a long time.

More precise measurements were then carried out with "vulcanized" rubber. The now solid, but finitely extensible, sample was measured under no load as a function of temperature. Its volume and heat capacity were determined. The thermal expansion coefficient was similar to that of a liquid, even though it was definitely a solid. After all the experiments with weights and temperature changes, the well annealed sample returned to its exact initial length; no permanent set was achieved. However, during some experiments at lower temperatures, crystallization led to hysteresis, and sluggish attainment of equilibrium was observed. The stress-strain curve was curved and attained a slope close to 0 when larger extensions were achieved. The additional heating effect also decreased since very little additional extension was achieved. The phenomenological Laws of rubber elasticity are one of the cornerstones of modern polymer science, even though they were established in the middle of the nineteenth century. Only the field of polymer science was missing.

Fundamental organic chemistry studies of isoprene and its polymers were initiated by Hermann Staudinger (1881–1965, Nobel 1953) at the Eidgenossische Technische Hochschule (ETH) in Zurich. He was able to synthesize polyisoprene. He proposed that it had a linear molecular structure composed of ordinary chemical bonds. One of the most illuminating studies of "Kautschuk" was the full hydrogenation of natural rubber by Staudinger and Fritschi [15]. The product was also rubbery and dissolved completely to give a very viscous solution. Staudinger declared that it was a true "Makromolekule!" In one of his seminal papers on natural rubber he defined the term in organic-structural terms:

For those colloidal particles in which the molecule is identical with the primary particle and in which the individual atoms of the colloidal particle are linked together by normal valences, we propose the term **macromolecules***. Such colloidal particles form true colloidal materials, which, in accordance to the bonding power of carbon, occur particularly in organic chemistry and in organic natural substances. Here the colloidal properties are determined by the structure and size of the molecule* [15] (Fig. 2.4).

Further progress in understanding natural rubber had to await the discovery of more microscopic probes of matter. Jean Perrin (1870–1945) established that ordinary matter is composed of atoms of the currently understood size. What probe is also of this size? X-rays or electrons! Soon after the experimental field of X-ray scattering had been developed, natural rubber was examined by Johan R. Katz (1880–1938) of Amsterdam [16]. Rubber at rest yielded only an amorphous halo, characteristic of a liquid. When the sample was stretched until it crystallized, sharp features appeared. This confirmed the notion of strain induced crystallization. But, the shock was that the unit cell for crystallized natural rubber was quite small. If

the unit cell was composed of several molecules of the substance in question, then caoutchouc was not a polymer at all, but some sort of molecular aggregate. The mystery was solved by Michael Polanyi (1891–1976), who realized that the unit cell is the repeating part of the structure. If the macromolecule itself has "repeat units," then parts of several chains can be contained in each unit cell, but only a few intramolecular repeat units. Herman Mark (1895–1992) and Kurt Meyer (1883–1952) published their magisterial book, "Der Aufbau der hochmolekularen organischen Naturstoffe" in 1930 [17]. Polymer science was now on the horizon, but there was not yet a coherent community of scientists committed to the macromolecular paradigm.

After a thorough preparation in X-ray scattering at the Kaiser Wilhelm Institute in Berlin with Polanyi, Herman Mark was invited to join I.G. Farbenindustrie by Kurt Meyer. Mark was appointed the Director of Research in a department called the Laboratory of Highmolecular Compounds in Ludwigshafen. He followed up the initial studies of Katz, and, with G.V. Susich, solved the detailed chemical structure of the natural rubber crystal. The isomeric structure of poly(1,4-isoprene) was *cis* along the chain, relative to the $C = C$ bond, for natural rubber. (Guttapercha is *trans.*). The scientific freedom granted to Mark by Meyer and I.G. Farben was invaluable in the founding of polymer science. In 1932 Herman Mark returned to Vienna as Professor of Chemistry. He developed a collaboration with Eugene Guth (1905–1990, a brilliant theoretician) and they formulated one of the first successful theories of the molecular basis of rubber elasticity [18]. The key concept was that the long chains in natural rubber are highly tortuous in their relaxed state. Stretching leads to orientation (and hence birefringence). This constraint greatly lowers the entropy of the system. It is the large negative entropy change associated with elongation that provides the strong restoring force. Polymer science had arrived! (Figs. 2.5 and 2.6)

Fig. 2.5 Herman Mark
(1895–1992) The Geheimrat
of Polymer Science

Fig. 2.6 Kurt Meyer (1883–
1952) Raised Polymer
Science to the ETH

2.2 Polystyrene (Presented to HIST, March 2011)

Caoutchouc was easily isolated from tree sap in its fully polymeric form. The next materia polymerica was accidently discovered when a standard ingredient in perfume was heated and allowed to sit: polystyrene.

Andrew Ure's "Dictionary of Chemistry and Minerology" of 1831 lists Storax as "the resinous exudate of the Sweetgum (Liquidambar) tree" [19]. The sweet fragrance is due to cinnamic acid. When pure cinnamic acid is heated in a steam distillation it readily decarboxylates to yield styrene.

An earlier citation in William Nicholson's "A Dictionary of Practical and Theoretical Chemistry" of 1808 describes experiments by Neuman where he subjected Storax resin to steam distillation and obtained an empyreumatic oil [20]. The first person to "name" the oil was E. Simon in 1839: he called it styrol [21]. Careful elemental analysis of this substance by G. Gerhardt and A. Cahours in 1841 yielded the empirical formula CH [22]. Vapor density measurements, a standard in the land of Regnault and Dumas, yielded the correct molecular formula: C_8H_8. Extensive physical measurements on styrol by E. Kopp in 1845 yielded its boiling point(144 C), and its specific gravity (0.928). Chemical analysis with bromine revealed a single aliphatic double bond [23].

After preparing styrol, Simon set it aside for a few months and was surprised to discover that it had become a jelly-like material. He attributed this to oxidation by the atmosphere and called it styrol oxide. Many "drying oils" are cured by this process, but elemental analysis of the gel by J. Blyth and A.W. Hofmann in 1845 yielded the same composition as pure styrol [24]. They called it metastyrol. They also carefully observed the temperature during a controlled heating of styrol. Once the gelling process started, considerable heat was given off by the reaction. The final product of this reaction was a transparent glass. (Some of the most beautiful optical samples of pure polystyrene have been prepared in my laboratory by carrying out this process entirely in a vacuum system and polymerizing the liquid at 90 C). Unlike isoprene, which required some sort of catalyst to initiate polymerization, styrene polymerizes spontaneously, even at room temperature.

One of the high points of the period before the advent of the structural theory of organic molecules, associated with Kekule and others, was the publication of Marcellin Berthelot's "Chimie Organique Fondee Sur La Synthese" in 1860 [13]. Berthelot carried out extensive studies of the polymerization of styrene. In addition to obtaining styrol from cinnamic acid, Berthelot invented a synthesis using benzene and ethylene flowing through a hot metal tube. Berthelot was one of the foremost chroniclers of organic compounds and published "Les Carbures d'Hydrogene" regularly from 1851 to 1901 [25]. He organized related substances into monomers and "polymers" if they had the same empirical formula. He considered metastyrol to be a cyclic trimer of styrol. The nature of the link between monomers was left obscure, but intramonomer structure was also unclear at this point in history. Berthelot carried out quantitative measurements of the heat of polymerization for styrene, along with thousands of other reactions. He quantified the highly exothermic nature of this reaction (Fig. 2.7).

Fig. 2.7 Marcellin Berthelot
(1827-1907) Pere des
Polymeres?

   With so much effort and not inconsiderable chemical insight, it might be supposed that Berthelot could have become the "Pere des Polymeres". The lack of a clear structural theory of molecules was certainly a hindrance. Philosophical reticence with regard to speculative chemistry was reinforced by the Comtean winds blowing in France and Germany. Only "Positive" facts could be mentioned [26]. The thermochemical studies were extensive and mostly accurate, but no sense of the importance of entropy (discussed in detail by Carnot!) ever dawned on Berthelot. He missed the chance to explain chemical equilibrium in correct detail. Other Frenchmen (Le Chatelier) did not miss their opportunity! No one at this time understood the Nature of the chemical bond, but even when the structural chemists started to distinguish different bond orders, Berthelot did not join them.

   The seminal moment in the history of structural organic chemistry was the publication of "La chimie dans l'espace" by J.H. van't Hoff (1852–1911) in 1875 [27]. He envisioned molecules as chemically bonded collections of atoms in full three dimensional space. He explained the phenomenon of optical activity in terms of the arrangement of the four different substituents of the tetravalent carbon along the four tetrahedral directions. The scientific community of Stereochemistry dates from this era. Van't Hoff was also a superb experimental chemist and corrected a serious misunderstanding promulgated by Berthelot with regard to the optical activity of styrene. Berthelot reported a measurable optical rotation of pure styrene, in violation of the principles elucidated by van't Hoff. The solution to this problem was a better preparation and purification of the styrene, whereupon no optical rotation was obtained. Van't Hoff also understood the stereochemistry of larger molecules and discussed the stereotacticity of oligomers of styrene [28]. If the backbone is viewed in its all-*trans* conformation, phenyl rings are connected to every other carbon. The side groups can be on the same or opposite sides of the backbone. All the structural concepts currently employed in polymer science are already contained in "Chemistry in Space." Van't Hoff went on to revolutionize Physical Chemistry and won the first Nobel Prize in Chemistry in 1901. Berthelot would have done well to listen to a Dutchman!

Polystyrene remained a laboratory curiosity until the mid twentieth century. The price and commercial availability of the monomer inhibited industrial development. The story of the eventual commercialization of polystyrene is beyond the scope of this chapter, but the motivation was its use as a component of synthetic rubber needed for the war effort. Styrene-butadiene rubber was the copolymer of choice.

Scientific study of polystyrene was taken up with vigor by Hermann Staudinger in the 1920s. The record is largely contained in his monograph "Die Hoch-molekularen Organischen Verbindungen" [29] and in the thesis of W. Heuer. Hundreds of samples of polystyrene were synthesized and fractionated. They were studied as a function of concentration and temperature using viscosity in a number of different solvents. The amount of data is impressive, sort of like Berthelot. Staudinger proposed that the data could be represented in the form of a universal law for the intrinsic viscosity:

$$[\eta] = \lim_{c \to 0}((\eta - \eta_0)/\eta_0 c) = KM$$

where K is a constant and M is the molecular weight. In fact, K depends on solvent and temperature. The clearest conclusion that can be drawn from the actual data reported in the monograph is that this law is not valid for all molecular weights, and that it is not valid for all solvents, and that it does not seem to be valid for any actual system. While Staudinger did contribute much to the study of polystyrene, he failed to understand its most characteristic properties. Like Berthelot, he should have listened to the Dutchman!

2.3 Bakelite (Presented to HIST, March 2010)

One of the most remarkable materials ever created by humans was synthesized from inexpensive industrial chemicals: Bakelite. It is composed of the reaction products of phenol and formaldehyde. Phenol (hydroxybenzene) (C_6H_5OH) was discovered as a component of coal tar in 1834. It was used for a variety of medicinal purposes throughout the nineteenth century. In the 1870s, Adolph von Baeyer (1835–1917) carried out systematic studies of the reactions of phenol with aldehydes, such as acetaldehyde and benzaldehyde [30]. The reactions were cat-alyzed by strong acids. The products were highly viscous liquids. Baeyer was one of the leading synthetic chemists of the nineteenth century and received the Nobel Prize in 1905.

Another leading chemist to study these systems was Arthur Michael (1853–1942). He demonstrated that the reactions of phenol and aldehydes could also be catalyzed by bases [31]. Michael was a member of the U.S. National Academy of Sciences (1889) and a Professor at Harvard University (1912).

By 1890, formaldehyde could be obtained commercially as a 40% aqueous solution. Research in the famous laboratory of Emil Fischer (1852–1919, Nobel 1902) was instituted by W. Kleeberg on the reactions of phenol and formaldehyde. When hydrochloric acid was used as the catalyst a resinous product was obtained. Heating of this substance produced a solid which was insoluble in most solvents [32].

A more detailed study of the reactions of phenol and formaldehyde was carried out by O. Manasse [33], and independently by L. Lederer [34]. A basic catalyst was employed. The initial product was analyzed and found to be saligenin (o-hydroxybenzyl alcohol). Further work also isolated p-hydroxybenzyl alcohol. When saligenin was reacted in the presence of acid, it oligomerized into saliretin, a mixture of many substances.

Our story takes an important turn when W.H. Story applied some of the methods of industrial chemistry to the reaction of phenol and formaldehyde [35]. He constructed a heated pressure reactor with inlet and outlet valves, and an independent source of gas pressure. The system was charged with commercial carbolic acid (phenol) and 40% formaldehyde solution. It was heated at 100 °C and stirred for 8 h. The highly viscous solution was then drawn off and concentrated to drive off water. Further heating led to a solid product that was clear, tough and a good electrical insulator. But, chemically, what was it?

Leo H. Baekeland (1863–1944) combined the fundamental knowledge of a good academic synthetic chemist with the thorough practical procedures of a great industrial chemist, like Hancock. He tried everything! First he verified the reported syntheses involving phenol and formaldehyde. He isolated saligenin and carried out reactions with excess phenol; diphenylol methanes were produced. He evaluated the products produced under both acidic and basic conditions. He concluded that acid catalysis produces oligomeric products that are now called Novolak resins. He varied the stoichiometry of ammonia as a basic catalyst and used only equimolar amounts (Fig. 2.8).

Baekeland described the reaction in terms of three phases: In the initial phase, a condensation reaction occurred. The water produced as a product was removed using a system like that described by Story. During the second heating stage, the viscous mass became a soft solid that could be swelled with an appropriate solvent. During the final heating phase under pressure, the system became a hard, insoluble mass. This transparent product can be heated to high temperatures without melting or charring. It is better than the best natural amber. The chemical analysis yielded a pure oxy-hydrocarbon with a formula $(C_{43}H_{38}O_7)_n$. This thrilling story was told by Baekeland to the New York Section of the American Chemical Society in 1909 and was published in the initial volume of the Journal of Industrial and Engineering Chemistry [36].

Now that he had a great product, he proceeded to optimize the industrial process. He determined that the best charge for the "Bakelizer" was in the ratio of 7 mol of formaldehyde to 6 mol of phenol. The initial products were ortho and

Fig. 2.8 Leo H. Baekeland (1863–1944) Inventor of Bakelite (Karraker, by permission)

para-hydroxybenzyl alcohol. He continued to carry out fundamental studies of this reaction throughout his industrial career and was rewarded with the Willard Gibbs Medal of the Chicago Section of the American Chemical Society in 1913 [37]. His acceptance speech emphasized the following points:

> *First: That these bodies are phenolic condensation products of formaldehyde, this condensation process having for result a corresponding enlargement of the so-called carbon nucleus of the molecule.*
>
> *Furthermore, it is a well accepted fact that in these reactions, formaldehyde can be replaced by its many equivalents.*
>
> *Second: That after the so-called condensation has taken place polymerization sets in, with the result that the molecules of condensation product form by aggregation or regrouping, so-called polymerized molecules of much higher molecule weight. This at once explains the higher specific gravity and contraction of the polymerized product.*
>
> *If we go beyond these mere general theoretical conceptions and try to interpret the intimate chemical constitution of these bodies in about the same way as we are able to do with the relatively much simpler crystalline or volatile organic compounds, then our flights of fancy are justified to take as well one direction as another. If beginners in organic chemistry may be impressed to a certain extent by the proposed formulae, which try to represent the molecular structure of these refractory bodies, the experienced organic chemist will only consider them as a matter of very subordinate interest, and merely as a crude attempt to show one of the many ways in which the constitution of these interesting products might be explained, according to our rather insecure theoretical notions* (Fig. 2.9).

In order to produce a deeper study of the fundamental organic chemistry, Baekeland supported a graduate fellowship at Columbia carried out by H.L. Bender. The published version of the thesis appeared in Industrial and Engineering Chemistry in 1925 [38]. The emphasis was on isolating and characterizing all the initial products of the reaction. Saligenin is crystallizable and was easily demonstrated. However, the mechanism of the formation of the benzyl alcohols was further elucidated by proposing phenoxymethylalcohol as the actual initial product. This substance undergoes rapid rearrangement to the benzyl

Fig. 2.9 Bakelizer
(Smithsonian Institution, by
permission)

alcohol. This is an example of a Claisen rearrangement. Either saligenin or the para-hydroxybenzylalcohol then reacts with phenol to produce phenoxy-p-hydroxyphenylmethane. This compound can, and occasionally does, rearrange to dihydroxyphenylmethane. It can be crystallized and was isolated by Baekeland. The mixed product remains in the amorphous state. The measured molecular weight of both substances was 200 Daltons (Fig. 2.10).

Baekeland then considered the further reaction in the presence of more formaldehyde and under basic conditions. Where could the reaction come from? He proposed that the formaldehyde attacked the methane carbon to produce an unsaturated molecule, a substituted ethylene. The vinyl monomer can then polymerize under the application of heat to produce high polymer. It is notable that in 1925 Baekeland is discussing seriously the polymerization of an organic compound. Work by Staudinger and by Carothers was only in its infancy. Perhaps the macromolecular paradigm was more widespread in the United States than in Europe!

A brilliant summary of the progress in the resin industry was published in 1926 by Barry, Drummond and Morrell [39]. A.A. Drummond recruited N.J.L. Megson of the British Government Department of Scientific and Industrial Research to study the detailed pathways of the phenol-formaldehyde reaction. The basic philosophy was that if the proposed intermediates could be synthesized by rational

Lilley first formulates the ionic mechanism thus:

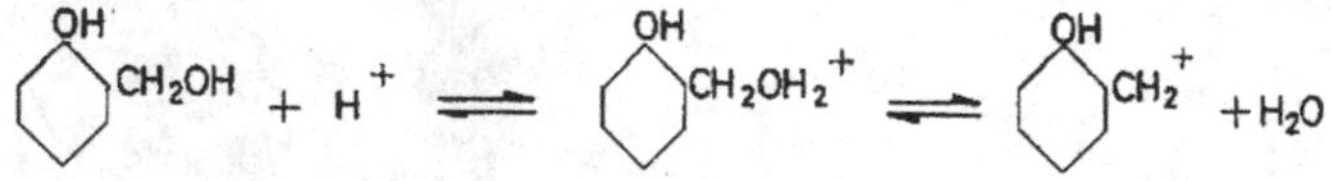

followed by the reactions 1 and 2 below:

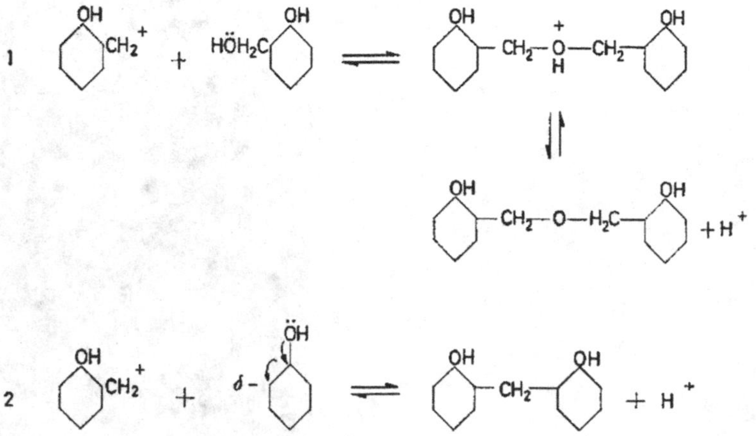

Fig. 2.10 Reaction scheme for saliretin and phenol (Megson, by permission)

organic methods, a clearer picture could be constructed. Megson reported his results in a brilliant lecture at the Faraday Discussion on Polymerization in 1935 [40]. He was able to construct a full three dimensional model of the local chemical structure of Bakelite (Fig. 2.11).

Van't Hoff would have been proud!

2.4 Polysaccharides

The earth is filled with polysaccharides. They have impacted human culture since the beginning. The growth in understanding of this class of macromolecules was slow, but the first Nobel Prize in Chemistry that was awarded for work that specifically included polymers was given to Sir Walter Norman Haworth (1883–1950, FRS) in 1937 for his work on polysaccharides. These materials constitute one of the most fascinating groups of macromolecules (Fig. 2.12).

Two of the most common polysaccharides are sugar and starch. When Chemistry had advanced enough to carry out elemental analyses with good precision, Gay-Lussac and Thenard discovered that these substances had the same

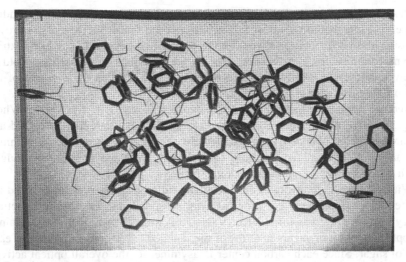

PLATE 3. *Model of phenolic resin (Reproduced by permission of Mr. E. G. K. Pritchett.)*

Fig. 2.11 Megson model of solid Bakelite (Megson, by permission)

Fig. 2.12 Sir Walter
Norman Haworth (1883–
1950, Nobel 1937) (Nobel
Archives)

empirical formula, CH_2O, and called them "carbohydrates" [41]. Kirchoff discovered that starch could be converted completely to sugar by acid hydrolysis [42]. Biot and Persoz were able to isolate intermediate products of the hydrolysis which they called "dextrins," because they exhibited a strong rotation of the plane of polarized light [43].

Dextrins were one of the first polymeric materials to be successfully studied with cryoscopic methods [44]. It might be supposed that from then on, improved precision and better materials would have established beyond doubt that dextrins were macromolecules. But, there are times when the winds of science blow cold on inconvenient "facts." Only approved results are often given the blessing of the high priests of science.

Another substance of ancient acknowledgement is wood fiber, cellulose. Whole samples of tree wood differ from species to species, but carefully extracted and prepared samples obtained from wood were shown by Payen to be a common substance which he called cellulose [45]. Fremy was able to show that sugar, starch and cellulose are all carbohydrates [46].

With so many carbohydrates, and so many different properties, how could the differences be explained? No progress was possible until the atomic structural theory of molecules was formulated. Van't Hoff went even further and explained the optical activity of sugars in terms of the detailed asymmetric structure of each kind of sugar. Since each carbon center is asymmetric, the overall optical activity depends on the detailed local structure of the sugar. But, what was this structure?

Cellulose can also be broken down into "mers" by the application of strong acid and the resulting product is entirely glucose, just like starch! The differences become immediately apparent in the dimer, trimer, etc. The dimer, cellobiose, was isolated by Franchimont in 1879 [47]. A whole series of oligomers were isolated by Willstatter and Zechmeister in 1929 [48]. Cellulose is a polymer of glucose. But, so is starch! The explanation required a detailed understanding of the chemical structure of glucose. This was provided by Haworth in 1927 [49]. There are actually two distinguishable forms of glucose that differ when they form the hexose ring mers observed in cellulose.

α - Glucose   β - Glucose

There can then be many different bioses. Cellulose is based on cellobiose, while starch is based on the biose, maltose.

β - Cellobiose   α - Maltose

The easiest way to see the difference is to focus on the two methoxy groups. They are on the same side in maltose, and on opposite sides in cellobiose. The

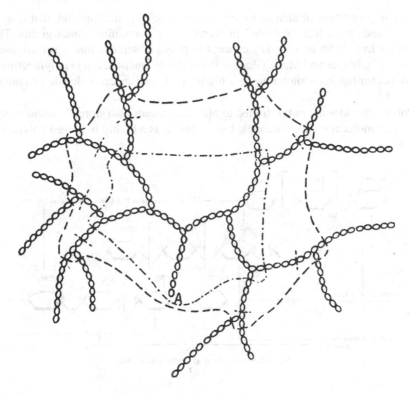

Fig. 2.13 Amylopectin Skeleton (Haworth, by permission)

detailed chemical structure of cellulose was determined by Meyer and Mark using X-ray diffraction from crystalline cellulose fibres [50].

A comparable representation of the amylose component of starch grains is:

Starch chain

Starch grains also contain an isomeric poly(glucose), amylopectin, that is quite variable and much less "soluble" in warm water containing starch grains. This substance introduces another key concept in polymer science that was recognized by Haworth, Meyer and Mark. Glucose is not just bifunctional, it is polyfunctional. Amylopectin has been shown to be a highly branched macromolecule of maltose (Fig. 2.13).

Polysaccharides are not restricted to plants. Animals also store chemical energy in a macromolecular substance called glycogen. It is a highly branched polymer of maltose.

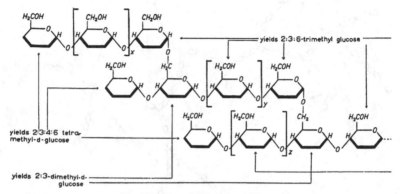

Fig. 122.—Formula of glycogen (Haworth).

The structural element in most fruits is a polysaccharide called pectin. While there may be many substances in a particular type of fruit, the dominant macromolecule is called polygalacturonic acid.

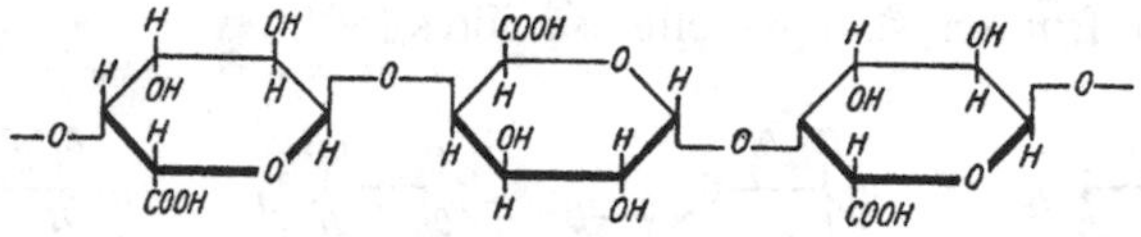

Fig. 110.—Formula of polygalacturonic acid from pectin
(Meyer and Mark).

While there may have been controversy in other quarters, Haworth, Meyer and Mark were doing polymer science. Careful experiments were interpreted in terms of detailed molecular models. All the best concepts of structural chemistry and all the best experimental methods from X-ray diffraction to detailed synthetic analyses were employed. Eventually a much larger community would join them.

References

1. Loadman J (2005) Tears of the tree. Oxford University Press, Oxford
2. Fresneau F, de la Condamine CM (1751) Memoires de l'Academie Royales de Sciences 319
3. Coleby LJM (1938) The chemical studies of P.J. Macquer. George Allen & Unwin Ltd, London
4. Macquer PJ (1763) Histoire de l'academie Royale de Sciences 49
5. Macquer PJ (1768) Memoires de l'Academie Royale de Sciences 209
6. Murray J (1806) A system of chemistry, Vol. 4. 4vols. Longman, Hurst, Rees and Orme, Edinburgh, p 180
7. Loadman J, James F (2010) The Hancocks of Marlborough. Oxford University Press, Oxford
8. Faraday M (1859) Experimental researches in chemistry and physics. Richard Taylor and William Francis, London
9. Ure A (1856) A dictionary of arts, manufactures, and mines. Vol. 1. 2 vols. D. Appleton and Company, New York
10. Williams CG (1860) On isoprene and caoutchine. Philos Trans Royal Soc Lond 150:241–255
11. Fisher HL (1957) Chemistry of natural and synthetic rubbers. Reinhold Publishing Corporation, New York
12. Tilden WA (1884) J Chem Soc 45:401
13. Berthelot M (1860) Chimie Organique Fondee Sur la Synthese. Mallet-Bachelier, Paris
14. Joule JP (1858) On some thermo-dynamic properties of solids. Philos Trans Royal Soc Lond 149:91–131
15. Staudinger H (1924) Uber die Konstitution des Kautschuks. Berichte der Deutchen Chemischen Gesellschaft 57:1203–1208
16. Katz JR (1925) Was sind die Ursachen der eigentumlichen Dehnbarkeit des Kautschuks? Kolloid Zeitschrift 36:300–307
17. Meyer KH, Mark H (1930) Der Aufbau der hochpolymeren organischen Naturstoffe. Akademische Verlagsgesellschaft, Leipzig
18. Guth E, Mark H (1934) Mh Chem 65:93
19. Ure A (1831) A dictionary of chemistry and minerology, 4th edn. Thomas Tegg, London
20. Nicholson W (1808) A dictionary of practical and theoretical chemistry. Richard Phillips, London
21. Simon E (1839) Annalen 31:265–277
22. Gerhardt C, Cahours A (1841) Annalen 38:67–108
23. Kopp E (1845) Compte Rendues des Academie de Sciences 21:1376–1381
24. Blyth J, Hoffman AW (1845) Annalen 53:292–329
25. Berthelot M (1851) Les Carbures d'Hydrogen. Paris
26. Rocke AJ (2001) Nationalizing science: Adolphe Wurtz and the battle for French chemistry. MIT Press, Cambridge
27. van't Hoff JH (1875) La Chimie dans l'espace. Utrecht
28. van't Hoff JH (1891) Chemistry in Space (trans: Marsh JE). Clarendon Press, Oxford
29. Staudinger H (1932) Die Hochmolekularen Organischen Verbindungen. Springer, Berlin
30. von Baeyer A (1872) Berichte der Deutchen Chemischen Gesellschaft 5:1094–1095
31. Michael A (1883) Am Chem J 5:338
32. Kleeberg W (1891) Annalen 263:263–283
33. Manasse O (1894) Ueber eine Synthese aromatischer Oxyalkohole. Berichte der Deutschen Chemischen Gesellschaft 27:2409–2413
34. Lederer L (1895) J Soc Chem Ind 14:297
35. Story WH (1905) J Soc Chem Ind 24:1081
36. Baekeland LH (1909) The synthesis, constitution, and uses of bakelite. J Ind Eng Chem 1:149–161
37. Baekeland LH (1913) The chemical constitution of resinous phenolic condensation products. J Ind Eng Chem 5:506–511

38. Baekeland LH, Bender HL (1925) Phenol resins and resinoids. Ind Eng Chem 17:225–237
39. Barry TH, Drummond AA, Morrell RS (1926) The chemistry of the natural and synthetic resins. D. van Nostrand Company, New York
40. Megson NJL (1958) Phenolic resin chemistry. Butterworths Scientific Publications, London
41. Gay-Lussac J, Thenard J (1811) Recherches Physico-Chimiques 2:268–350
42. Kirchoff C (1815) Journal de Chemie 14:385–395
43. Biot JB, Persoz JF (1833) Annales de Chimie Physique 53:73
44. Brown HT, Morris GH (1889) J Chem Soc 55:462
45. Payen A (1838) Memoire sur la composition du tissue proper des plantes et du ligneux. Comptes Rendues 7:1052–1056
46. Fremy E (1859) Compte Rendue des Academie de Science 48:361
47. Franchimont APN (1879) Ueber Kohlehydrate. Berichte der Deutschen Chemischen Gesellscahft 12:1938–1942
48. Willstatter R, Zechmeister L (1929) Berichte der Deutchen Chemischen Gesellscahft 62:722
49. Haworth WN (1929) The constitution of sugars. Arnold, London
50. Meyer KH, Mark H (1928) Berichte der Deutschen Chemischen Gesellschaft 61:593–614

Chapter 3
The Faraday Society and the Birth of Polymer Science*

Another lens through which to view the birth of polymer science is the foremost community of physical chemists in the early twentieth century, The Faraday Society [1]. It was founded in 1903 to promote "the study of Electrochemistry, Electrometallurgy, Chemical Physics, Metallography and kindred subjects." This definition appears in all the early issues of the Transactions of the Faraday Society. A further clarification is contained in the Council Report for 1903–1904: "the sciences included are those branches of pure and applied physical chemistry which do not come precisely within the scope of existing scientific and technical Societies. It is worthy of note that the Faraday Society is unique in being the only scientific body in England which specifically aims at encouraging and combining both the theoretical and practical sides of the subjects that come within its scope" [1]. This concern with industrially important topics allowed the Faraday Society to consider many subjects that were complex and relevant to both life and industry, such as electrochemistry and colloid science.

3.1 Discussion Meeting of 1907 on Osmotic Pressure

One of the signature activities of the Faraday Society was the public Discussion Meeting on controversial and highly relevant topics. The first such meeting was held on January 29, 1907 on the subject of "Osmotic Pressure." The results were published in Volume 3 of the Transactions of the Faraday Society. The theoretical ideas of van't Hoff [2] and the experimental developments of Pfeffer [3] had provided a sound basis for the active development of the experimental and theoretical field of Osmotic Pressure. The meeting was started with a presentation and

---

*Presented to HIST, August 2010

G. Patterson, *A Prehistory of Polymer Science*, SpringerBriefs in History of Chemistry, 25
DOI: 10.1007/978-3-642-21637-4_3, © The Author(s) 2012

Fig. 3.1 T. Martin Lowry (1874–1936, FRS, President Faraday Society) (Royal Society of Chemistry, by permission)

discussion of the current state of the art in experimental measurements by the Earl of Berkeley (1865–1942) and his assistant, E.G.J. Hartley. They demonstrated a new "vapor pressure osmometer." T. Martin Lowry (1874–1936, FRS) presented an overall introduction to the subject of "Osmotic Pressure from the Standpoint of the Kinetic Theory." The basic principles were explained in terms of the work of van't Hoff and Walther Nernst (1864–1941, Nobel 1920). These plenary lectures established that osmotic pressure had achieved paradigm status in the physical chemistry community (Fig. 3.1). However, later workers often tried to evade the results of osmotic pressure measurements and to suggest that "colloidal solutions" were not solutions at all and had no osmotic pressure. What was not settled science was the nature of solutions themselves. The American physical chemist from Wisconsin, Louis Kahlenberg, argued against Arrhenius, van't Hoff and Ostwald, that salts in solution were not ionized. Even at Faraday Society Meetings, the quality of the "science" was variable. What was encouraged was discussion by the best-known people in the field of interest.

As was to be the tradition, the liveliest part of the Meeting was the Discussion. Professor Alexander Findlay (1874–1966) of Aberdeen gave a stirring sermon on the difference between asymptotic "Laws" and the results for actual solutions at finite concentration. Findlay went on the publish a full book on Osmotic Pressure in 1913 [4]. Many substances in solution gave small osmotic pressures, indicative of large particle weight.

The argument then became one of identifying the nature of the osmotic particle. Some people, like Wolfgang Ostwald, insisted that all colloids were aggregates. This type of obfuscation hindered consensus in this area. It would take more than just a high particle weight to convince the skeptics that large molecules existed in solution.

3.2 Discussion Meeting of 1913 on Colloids and Their Viscosity

The next Discussion Meeting that pointed towards polymer science was held March 12, 1913 on the subject of "Colloids and Their Viscosity." It was published in Volume 9 of the Transactions of the Faraday Society. Einstein had written a definitive treatise on the viscosity of particles in solution. The intrinsic viscosity depended on the hydrodynamic volume of the particles. Measuring the solution viscosity was straightforward. Emil Hatschek (1869–1944) of the Sir John Cass Institute was in the Chair. The meeting commenced with a rambling romp through the swamps of colloid science by the leading huckster of the "New Science of Colloids," Wolfgang Ostwald (1883–1943). He urged the assembled worthies to consider the Ostwald viscometer as a true "Colloidoscope!" He emphasized that colloidal systems were characterized by unique and unexplained principles, and would provide a lifetime of complex and interesting problems. Like a river, no colloidal system would ever be the same twice! Some clarity was returned to the discussion by Victor Henri of Paris who delivered a synoptic presentation on "The Determination of the Size of Colloidal Particles." He reviewed methods such as viscosity, the ultramicroscope, the equilibrium concentration distribution in a gravitational field, the settling rate of particles in a gravitational field, diffusion, Brownian motion and Rayleigh scattering. These fields were based on the foundational theoretical and experimental work of Einstein, Smoluchowski, Perrin, Hatschek, Mie and Rayleigh. The best scientists in the world had provided a sound basis for examining particle size. Another clear and stimulating lecture was given by Wolfgang Pauli (1900–1958) of Vienna on protein solutions. He carried out studies of albumin and gluten as a function of salt concentration and pH. The concept of globular proteins in aqueous solutions was explained in detail that would be acceptable today. Another exceedingly careful study of aluminum hydroxide sols was reported by Professor H. Freundlich (1880–1941, FRS) of Berlin. Precise science was possible for colloidal systems, but the work also required a scientist, not a carnival showman. Colloid science also included studies of emulsions. Emil Hatschek presented a paper on "The General Theory of Viscosity of Two-phase Systems." Solution viscosity was an important tool in the eventual consolidation of polymer science. The theory and experimental procedures were already in good shape in 1913, but a knowledge of the hydrodynamic volume of the particle, like a knowledge of its particle weight, was not enough to convince Wolfgang Ostwald that molecules were involved. Snide remarks about "molecules" still haunt discourse in certain European circles.

3.3 Discussion Meeting of 1920 on Colloids

The War years halted most scientific activity in England, and no Discussion Meetings were held until 1919. Emil Hatschek had not forgotten the 1913 Meeting and proposed that a more general meeting on "The Physics and Chemistry of

Fig. 3.2 Theodor Svedberg
(1884–1971, Nobel 1926)
(Nobel Archives, by
permission)

Colloids" be held. It occurred on October 25, 1920 and was published in
Volume 16. It was co-sponsored by the Physical Society of London, and Sir
William Bragg (1862–1942, K.B.E., F.R.S.) was the Chair. The plenary lecture
was given by one of the rising stars in physical chemistry, The Svedberg
(1884–1971) of Upsala, who would receive the Nobel Prize in Chemistry in 1926.
He focused on the microstructure of matter at the molecular, mesoscopic(colloidal)
and macroscopic length scales. Due reverence was expressed with regard to
Graham and Faraday. The breadth and depth of his knowledge was vast, and he set
a high tone for the Meeting. He was familiar with everything that was going on in
colloid science. He knew the theoretical work of Einstein and Smoluchowski. He
knew the experimental work of Perrin and Zsigmondy. This brilliant synoptic view
provided the essential paradigm for the whole meeting (Fig. 3.2).

Wolfgang Pauli was back with another report on the careful synthesis and
characterization of soluble metal-oxide colloids. His lecture was entitled "The
General Structure of Colloids." While Pauli moved on to other fields where he
succeeded brilliantly (Nobel, 1945), his contributions to colloid science were
important and established high standards.

Another exemplar was F.G. Donnan (1870–1956, FRS) of University College,
London. He gave a plenary lecture on Emulsions. He outlined the concepts that
still define the field today: definition of components in a multicomponent system, the importance of surface energies, the wide array of solution morpholo-
gies, the importance of ionic strength in aqueous solutions, the role of
surfactants. He also included an exhaustive bibliography in the published version (Fig. 3.3).

Fig. 3.3 F.G. Donnan
(1870–1956, FRS, President
Faraday Society) (Royal
Society of Chemistry, by
permission)

The Discussion was marked by sparkling comments by Hatschek, W.C. McChesney Lewis and The Svedberg. The existence of a truly sound paradigm made discussion both possible and profitable.

One of the most dramatic phenomena in colloid science is the sol-gel transition. Emil Hatschek gave the introductory lecture on the "Properties of Elastic Gels." Elastic gels are locally more like a solution, and small molecule solutes have normal diffusion coefficients. Hatschek identified two key issues that needed increased effort: (1) elucidating the structure of gels, and (2) explaining the swelling behavior. Professor H. R. Proctor of Leeds gave an immediate answer: A gel is like a semi-permeable membrane and hence swelling is due to the Donnan equilibrium. Since the ions must have the same chemical potential inside as well as outside the gel in the surrounding solution, the concentrations must be different. This insightful thermodynamic analysis is correct. Even though Proctor did not yet know the details of the gel structure, he considered the swollen gel to be a homogeneous phase, and hence he could trust thermodynamics to make sound predictions. Sound science is based on good experimental work and irrefutable theoretical concepts such as thermodynamics. Attempts to obfuscate physical reality by appealing to "special" effects is not the way forward. The field of colloid science chose to follow people like Hatschek, not hucksters like Wolfgang Ostwald.

Another colloidal phase transition is the thermo-reversible gelation. S.C. Bradford of the Science Museum of London presented a paper on gelatine gels. He measured the molecular weight of the individual gelatine molecules and assigned values in excess of 10,000 D. The gels were formed by cooling from homogeneous solutions. Heating revealed a sharp melting point and Bradford attributed this effect to real crystallization of the gelatine molecules. This work is clearly polymer science as we know it now. Although he could not be present in person, Richard

Fig. 3.4 Sir Robert
Robertson (1886–1975, FRS,
KBE, Nobel 1947) (Royal
Society of Chemistry, by
permission)

Zsigmondy (1865–1938, Nobel, 1925) of Gottingen was invited to respond to the papers on gels. While he acknowledged that some gels formed by crystallization processes, he cautioned that this would not be the only path to gels. The colloid science community was permeated by first rate scientists who both provided good examples and insisted on high quality work. The section on gels was concluded with a masterful summary of the Discussion by Hatschek.

While aqueous solutions of colloidal particles are very important, there are also industrially important non-aqueous colloidal systems. The keynote speaker for this session was Sir Robert Robinson, (1886–1975, K.B.E., F.R.S., Nobel 1947), on "Nitrocellulose." Now this is a polymer! It is also a material of great military significance (hence the KBE). Raw cellulose undergoes many treatments before it is nitrated. The degree of polymerization of the "finished" cellulose is from 100 to 300 subunits. The prepared cellulose is then nitrated to varying degrees. A dry film of highly nitrated cellulose is a brittle, glassy material. The talk focused on the use of appropriate solution viscosity in the quality control of the manufacture of "gun cotton", but hints of polymer science can be heard in the sounds of battle. The following talk by F. Sproxton discussed solution fractionation of the nitrocellulose into samples of widely differing intrinsic viscosity. He correctly interpreted this as due to the inherent polydispersity of the initial processed cellulose. The concept of a distribution of molecular weights in many polymer samples is clearly present in this colloid community in 1920. The use of both intrinsic viscosity and osmometry was also standard by 1920 (Fig. 3.4).

Another industrially important material was natural rubber. B.D. Porritt (1884–1940) of University College, London presented a paper on the "Action of Light and Oxygen on Rubber." He discussed additives designed to absorb the light before it was absorbed by the rubber molecules, leading to scission. Chain scission was also produced by oxygen, probably through ozonolysis. Deliberate degradation of the chain length was produced by mechanical kneading (mastication). This was assessed by intrinsic viscosity. In 1920, natural rubber was a typical colloidal system!

One of the largest Discussion sections was devoted to precipitation. Many colloidal particles are held in solution by mutual repulsion between the electrically charged bodies. If the electrical double layer is reduced by increasing the ionic strength of the solution, the particles can approach one another close enough for van der Waals forces to lead to irreversible attraction and precipitation. This was a problem that was treated by both Zsigmondy and by Smoluchowski. This issue is still a hot topic in the colloid science community.

One of the most important classes of charged colloids is the globular proteins. Jacques Loeb (1859–1924) of the Rockefeller Institute presented a paper on "Proteins and Colloid Chemistry." He stressed the importance of the isoelectric point, since globular proteins had no net charge at this pH and hence were likely to precipitate. The science of polyelectrolyte systems remains a challenging topic today.

The breadth of the Colloid Science community in 1920 was very large. The industrial importance of colloidal systems provided a source of funds for research. Most of the key concepts that define the colloidal paradigm were already in place in 1920. Polymers fit well into this community, and the fundamental ideas were helpful in developing a picture of macromolecules in solution and the bulk. While some people, like Wolfgang Ostwald, rejected the notion of large molecules in favor of vague association complexes, the leaders of this community were quite favorable to the recognition of certain substances as polymeric and molecular. The Faraday Society was an equal opportunity club where all good ideas were welcomed, and even controversial ideas could get a hearing and a thorough discussion.

3.4 Cambridge Colloid Discussion of 1930

The Faraday Society decided in 1930 to form a Colloid Committee composed of representatives from the Royal Society, the Biochemical, Chemical, Physical and Physiological Societies, the Society of the Chemical Industry and itself. The first of three meetings planned by the Colloid Committee was held at Cambridge on September 29,1930 on the subject of "Colloid Science Applied to Biology." It was suggested by Sir William Bate Hardy (FRS). It attracted a very large crowd, and had to be moved to larger quarters at Cambridge. There were some truly notable talks, but the spirit of polymer science was not in the air at this Meeting. The talks were published in Volume 26 of the Transactions of the Faraday Society (1930).

As would be appropriate for a plenary lecture on biology, the Dean of Cambridge biologists, A.V. Hill, discussed "Membranes." Biological membranes were presented as highly dynamic objects in steady state. Wolfgang Pauli discussed the

Fig. 3.5 Sir Robert Mond
(1867–1938, FRS, President
Faraday Society) (Royal
Society of Chemistry, by
permission)

physical chemistry of protein solutions with his usual rigor and clarity. The Svedberg presented a mountain of data on electrophoresis and ultracentrifugation of proteins. Leslie J. Harris of Cambridge presented conclusive proof of the zwitterionic character of proteins.

A hint of polymer science was present in the discussion of the structure of cotton fibres. F.T. Pierce of the Shirley Institute in Manchester discussed X-ray scattering from cotton fibres. Further discussion of this topic was presented by O.L. Sponsler of the University of California at Los Angeles. J.T. Edsall of Harvard reported on his studies of myosin fibres in muscle. Colloid science is central to a discussion of the cell. Polymer science is also absolutely necessary, and eventually both fields were acknowledged as part of the multidisciplinary study of living systems.

3.5 Discussion Meeting of 1932 on the Colloid Aspects of Textile Materials

The second large Discussion Meeting planned by the Colloid Committee was held at the University of Manchester from September 21–23. Sir Robert Mond (1867–1938, FRS), the President of the Faraday Society, was in the Chair. The papers were published in Volume 29 (Fig. 3.5).

After a few brief introductory remarks by F.G. Donnan, the initial plenary lecture was given by Herman Mark of Vienna. This was a real coming out party for Mark in England. His work was known and appreciated in Manchester, but by 1932 it had reached the point where historic advances were being made. This could have been the occasion for the birth of polymer science as we know it, but as we shall see, not everything was yet auspicious for this event. Mark's technical talk was on studies of stress and strain and the resulting orientation of cellulose micelles in cotton fibres. The fundamental cellulose valence chains were from 150 to 200 subunits long. These are indeed macromolecules.

Another center of textile fibre science was Birmingham and Sir Walter Norman Haworth (1883–1950, FRS, Nobel 1937) presented a very detailed report on the primary valence chains of several polysaccharides. Cellulose was shown by direct chemical analysis to contain from 100 to 200 subunits in a linear chain. Starch was analyzed in a similar fashion and found to be from 24 to 30 glucopyranose units. Glycogen is even shorter(12 units). The fundamentally different substance, inulin, is a fructofuranose of approximately 30 units. Macromolecules are envisioned in a fully modern sense in this paper. The three dimensional aspects were also addressed, since 1,4-glucose linkages produced different chains than 1,2-furanose linkages.

The paper by Hermann Staudinger was also highly anticipated: "Viscosity Investigations for the Examination of the Constitution of Natural Products of High Molecular Weight and of Rubber and Cellulose." Polymer solutions can display very large shear viscosities, even at very low concentrations. Staudinger attributed this to the presence of macromolecules at a time when many others still thought the particles were colloidal aggregates. His seminal monograph, "Die Hochmolekularen Organischen Verbindungen", appeared in 1932 [5].

The problem with his views was that they were supported with bad data and bad arguments. He asserted that macromolecules in solution retained the same conformation as they had in the crystalline state: extended helical rods. He claimed that his viscosity measurements entailed this conclusion. He also claimed that the so-called intrinsic viscosity, $[\eta] = \lim_{c \to 0} ((\eta/\eta_0) - 1)/c = KM$, was proportional to the first power of the molecular weight M. (K is a universal constant). The discussion of this paper was especially acrimonious. Herman Mark led the attack, but he was supported by Hayworth, Herzog, Adam, Rideal, Katz and Neale.

The extensive General Discussion revealed the heterogeneous nature of the "fibre" community. There were macroscopic textile fibre scientists who treated their systems at the whole fibre level (Shirley Institute). There were colloid scientists who mostly argued about nomenclature. Interestingly, Staudinger was from this group. There were technologists who studied actual industrial materials of heterogeneous character (never the same twice). And there were actual molecular scientists who focused on the details at the molecular level (Mark, Haworth, Rideal, Katz, Kratky, Hirst, etc.)

The study of polymeric systems was alive and well at the Kaiser Wilhelm Institute in Berlin. R.O. Herzog, O. Kratky and E. Petertil presented their most recent work on light scattering from polymer solutions. They were able to use the Gans theory of light scattering to provide some context for their results. They obtained molecular weights for various polystyrenes from 4000 to 440,000 D. Within a few years, light scattering from polymer solutions became one of the pillars of the foundation of polymer science.

Fig. 3.6 W.T. Astbury
(1898–1961, FRS)

This Meeting was also a coming out party for W.T. Astbury (1898–1961, FRS) of Leeds. He studied with Bragg and became one of the most famous X-ray diffractionists in the world, specializing in proteins. His talk at this meeting was on wool fibres (Fig. 3.6).

Another luminary appeared at this Meeting: N.K. Adam (FRS) of University College, London. Along with Irving Langmuir, he was the foremost experimentalist in the study of thin films. He reported on Langmuir–Blodgett films of cellulose derivatives. His classic book, "The Physics and Chemistry of Surfaces", appeared in 1930. He was a molecular physicist in the modern sense. During the Discussion, Mark immediately recognized the significance of this work for the understanding of the local structure and flexibility of macromolecules: they were flexible and dynamic, not rigid and straight!

F.D. Miles of the Nobel Laboratories declared: "The long chain molecule is now accepted as the best explanation of the chemical and physical properties of cellulose and its derivatives." However, there are times when the sociology of science lags behind the technical advances. The "gel point" for polymer science was only a few years away, but it was clear from such sentiments that the crosslinking reaction was well on its way in the human sol.

The understanding of X-ray scattering from condensed media was already advanced in 1932. J.R. Katz of Amsterdam reported studies of the X-ray pattern as a function of stretching of natural rubber. In the relaxed state, the pattern is characteristic of a simple liquid. At low levels of stretching, the sample becomes birefringent and the X-ray pattern is no longer isotropic. With sufficient stretching a sharp pattern emerges, suggestive of crystallinity. These changes are fully reversible. Astbury immediately grasped the significance of this work for the understanding of conformation changes during stretching: the highly contorted initial macromolecules had to unfold to crystallize.

Lord A.H. Hughes (FRS) of Cambridge summarized the state of understanding of protein structure: "The work of Astbury and Meyer and Mark provides ample evidence in support of the peptide chain theory of protein structure." But ample evidence is seldom enough to convince an entire community, such as the biological community, of the error of its position.

Happily, the Meeting closed with an appeal from Staudinger for cooperation between synthetic chemists, colloid chemists and physical chemists in their pursuit of an understanding of fibrous materials.

3.6 The General Discussion on Colloidal Electrolytes in 1934

The last of the three meetings planned by the Colloid Committee was held at Unversity College, London from September 27–29. Professor F.G. Donnan of UCL was the Chairman. The papers were published in Volume 31 (1935).

The Introductory lecture was given by Professor H. Freundlich, now of University College, London. Micelles composed of amphiphilic molecules form a bewildering array of structures as a function of soap concentration, pH and salt concentration. The highly dynamic nature of micelles was stressed.

One of the highlights of the Meeting was the presentation by G.S. Hartley of UCL on the Debye–Huckel Theory of Colloidal Electrolytes. This paper is still the foundation of current discussions on this topic.

Since proteins are colloidal electrolytes, they also played a pivotal role at this Meeting. F.G. Donnan discussed measurements of the electrovalency and osmotic pressure of protein solutions. G.S. Adair of Cambridge applied the theories of J. Willard Gibbs [6] to protein systems.

Proteins diffuse in solution and move in an electric field. Sir Erik K. Rideal (1890–1974, FRS) from Cambridge presented a very thorough discussion of these phenomena. Fellow workers in this area included G.S. Hartley from UCL and G.S. and M.E. Adair from Cambridge (Fig. 3.7).

The scientific research community of protein scientists could have provided one of the natural homes for polymer science, but the range of paradigms still rampant at that time was large, and the community became rather insular with respect to the more general communities of chemists and physicists. While outstanding work at the Nobel level was going on in this community, such as the work by Dorothy Jordan-Lloyd on leather proteins, there were always many Dorothy Wrinches in the group!

Fig. 3.7 Sir Eric K. Rideal
(FRS, President Faraday
Society) (Royal Society of
Chemistry, by permission)

3.7 Cambridge Meeting of 1935 on Phenomena of Polymerization and Condensation

Although the planned three colloid Discussions were now over, Lowry and Donnan seized the moment to suggest a Discussion of the rapidly developing field of polymer science. The response to this suggestion was dramatic. People came from all over the world to be present at the "birth of polymer science." It was held from September 26–28. The papers appeared in Volume 32 (1936).

The Discussion was Chaired by the current President of the Faraday Society, Sir W. Rintoul (O.B.E.). There was a special dinner before the meeting hosted by Rintoul and Donnan. The guests of honor were Hermann Staudinger, Kurt Meyer, Herman Mark, J.R. Katz and Wallace Carothers. It was a complete success (Fig. 3.8).

The Introductory address was given by Sir Eric Rideal (1890–1974, FRS). It was a masterful summary of the intellectual landscape to be explored at the Meeting. The protocol for these Discussions included presubmission of the papers so that all participants could consider them at length before arriving. This resulted in informed discussion. Rideal used the preprints to craft an essay that is still worth reading in the present!

This Meeting was a wonderful assemblage of academic, industrial, government and private scientists. The Phillips Laboratory at Eindhoven was well represented by J.H. deBoer and R. Houwink. deBoer gave a thorough discussion of the meaning of a molecule in the condensed state. He referenced two of the foundational monographs of polymer science by Meyer and Mark, "Der Aufbau der Hochpolymeren Organischen Naturstoffe" [7] and by Staudinger [5].

Fig. 3.8 Sir William Rintoul
(OBE, President Faraday
Society) (with permission
from Trans. Faraday Soc.
(1936)32:1485–1486)

The main course was served early. Wallace Carothers (1896–1937) of DuPont delivered a tour de force on "Polymers and Polyfunctionality." He stressed the need for the formation of actual chemical bonds between polyfunctional molecules. One class of bond forming processes is the condensation reaction. Carothers had made a thorough study of many polycondensation reactions. These reactions are reversible and lead to a product with a distribution of chain lengths. Carothers announced that the observed molecular weight distribution had been calculated by Paul Flory (1910–1985, Nobel 1974) from his group at DuPont. The discussion of this paper was very lively and Carothers gave especially good answers (Fig. 3.9)!

Another world center of polymer science in 1935 was Vienna. Herman Mark followed up his 1932 book on "Chemie und Physik der Cellulose" [8] with a summary of his current thinking on "The Mechanism of Polymerization." He emphasized that the existence of long chain molecules was now a scientific "fact," beyond further discussion. He illustrated the importance of studying the kinetics of the polymerization reaction to gain insight into its mechanism. His colleague at Vienna was Michael Polanyi (1891–1976), one of the world leaders in reaction kinetics. Mark explained the importance of initiation and growth processes, and identified termination, chain transfer and chain breaking reaction mechanisms. He also made reference to another theoretical colleague, Werner Kuhn (1899–1963) (Fig. 3.10).

J.R. Katz gave a Plenary lecture on "X-ray Spectroscopy of Polymers." He was a recognized expert on X-ray diffraction from crystalline materials and had trained with Bragg. He first heard Staudinger talk about macromolecules in 1927, but was unconvinced by his arguments. One of the major stumbling blocks for the X-ray community was that crystalline polymers displayed small unit cells. The solution

Fig. 3.9 Wallace Carothers
(1896–1937) Father of Nylon

Fig. 3.10 Michael Polanyi
(1891–1976) Physical
Chemist Extraordinaire

was discovered by Michael Polanyi: the unit cell reflects the size of the "repeating unit" of the crystal. Since polymers were themselves composed of repeating units, the unit cell was composed of a small number of mers from a small number of chains. The overall chain length could not be determined from the details of the unit cell. Katz also referenced the work of Houwink at Phillips. The discussion was vigorous but profitable. Meyer and Mark challenged Katz to help articulate and refine the understanding of X-ray scattering from polymers.

Hermann Staudinger presented his latest work on the "Formation of High Polymers of Unsaturated Substances." The amount of synthetic work was monumental. He stressed the importance of the "chain reaction" mechanism in these monomers. But, he could not resist the temptation to overgeneralize and to press doubtful points of science and nomenclature. The discussion was vigorous, but not very profitable. Kurt Meyer took the lead in this attack. Mark was more reserved but also more precise in his questions and comments. Fortunately, at this Discussion, antipathy between competitors was not able to derail the general goodwill. Although Staudinger received the Nobel Prize in 1953, he did not achieve an intellectual leadership role in the community of polymer scientists. His position was more that of a prophet.

The leading expert on the mechanical properties of bulk polymers, R. Houwink of Phillips, presented a paper on "Strength and Modulus of Elasticity of Some Amorphous Materials." Houwink also published a series of books on mechanical properties of polymers. The interchange in the discussion between Mark and Houwink emphasized the need for polymer scientists to be at the forefront of thinking in their area of expertise. Even if the experiments were well done, there is a need for a consistent conceptual scheme to interpret them. The discussion also heralded the appearance of N.J.L. Megson of the British Government Laboratory at Teddington. He discussed the structure and properties of phenol-formaldehyde resins.

The time then came for Kurt H. Meyer of Geneva to shine in his own light. He presented a paper on "Inorganic Substances with Rubberlike Properties." When crystalline sulfur is melted, its viscosity increases dramatically over time and eventually reaches an equilibrium state of very high viscosity. The viscosity changes with temperature, but is quite stable under anaerobic conditions. The thermodynamic heat of polymerization is positive for S_8, so that it is the entropy of the chain molecule that drives the reaction. This means that the chains have a very large number of conformations in the melt.

The principles of chemical kinetics were very important for the development of polymer science. Professor C.E.H. Bawn of Manchester presented a summary of the "Kinetics of Polymerization." The level of detail both for the reaction mechanism and the thermodynamics was very high. Bawn seemed to know every current paper in the field. Michael Polanyi had been driven out of Europe and was at Manchester in 1935. This collaboration was very fruitful. Bawn's paper provided the terms of discussion and a progressive research programme in polymer kinetics for the rest of the twentieth century (Fig. 3.11)!

Although many polymerizations take place in the bulk liquid or in solution, it is possible to polymerize monomers in a Langmuir–Blodgett film. Geoffrey Gee (FRS) presented his work carried out with Rideal at Cambridge. Polymerization even takes place in the gas phase, either homogeneously or on the wall of the vessel. The strong research group at Bristol was represented by Morris W. Travers who discussed new polymers of acetaldehyde. Sir Harry W. Melville (1908–2000, FRS) from Cambridge gave the latest results on mercury photosensitized reactions of acetylene. The high level of kinetic and mechanistic detail made polymeric systems one of the hot areas in chemical kinetics in England in the 1930s.

Fig. 3.11 C.E.H. Bawn
(President Faraday Society)
(Royal Society of Chemistry,
by permission)

Professor R.G.W. Norrish (1897–1978, FRS, Nobel 1967) at Cambridge also took an interest in the kinetics of polymerization.

Polymer science had been recognized in Moscow as a profitable area for research since the late 1800s. A paper on the kinetics of polymerization of 1,3-butadiene was presented by S. Medvedev. The conceptual foundations of polymerization had been established in Russia by N.N. Semenov [9].

While the shape of polymer molecules in solution had been a sore point of discussion in 1932, by 1935 there was increasing data and theory to compel a common paradigm. This was presented by R. Signer of Bern: "The Molecular Weight of Polystyrenes and the Shape of the Molecules in Solution." This experimental work was anchored in the theoretical work of W. Kuhn and the collaboration of Eugene Guth and Herman Mark. The so-called Staudinger viscosity Law was debunked both experimentally and theoretically. The consilience of great theory and sound experiments is one of the signs of a progressive research programme. Polymer science was clearly now in this category. The use of the "random walk model" for the conformation of flexible polymer molecules in solution was now under discussion and it has never left!

The picture of the attendees at this Discussion includes more than 200 scientists. Many of the speakers and discussion leaders became the pillars of the community of polymer scientists. In addition to the many people who devoted their primary efforts to polymers, there was a large group of leaders in the broader chemistry and physics communities that were interested in the progress in the understanding of macromolecules and constantly provided general guidance on matters of fundamental science. All the elements of a true paradigm driven community were now in place for polymer science (Fig. 3.12).

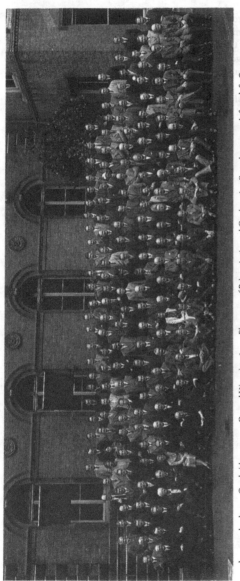

Group photograph taken at Cambridge, 1935 at General Dicussion 63 on *Phenomena of Polymerisation and Condensation*. Seated on ground, from left: Garner – – – – – – – – – – D. G. Davies – Allsop. First seated row, from left: T. C. James, W. H. Mills – – – – – – Lowry, W. H. Carothers, E. C. C. Baly – – – E. Mond, Standinger – – – – Sidgwick, Rideal, Mark, Lennard-Jones, de Boer(?), Houwink(?) – – – – First row standing, from left: – – S. R. Carter, G. Cox, Astbury – – – – – Dunstan, Cadman, Norrish, Travers, C. R. Bury – – Goodeve, C. Le Fevre, Mrs Goodeve – Marlow – – – – – – E. C. Bailey – Einich. Second row standing, from left: – – – – – – Salomon – R. Le Fevre – – – – – – Hartley – – – – J. K. Roberts – Allmand – Wansborough-Jones. Third row standing, from left: – – D. Crowfoot, Megaw, Bawn – – Boys(?) – – – – F. E. T. Kingman – – – – Bryce, Spence – – N. K. Adam – – F. I. G. Rawlins(?) –. Back row, from left: – – – – – – – – – – F. C. Frank, M. M. Davies, J. B. Marsden – – Mochwyn-Hughes, Gee, A. E. Alexander, Wynne-Jones – – –. Individuals deduced as being present at the meeting (from the report in the *Transactions*) but not identified: H. W Melville (Cambs), J. E. Carruthers (Cambs), S. C. Gran (Cambs), F. C. Wood (Manchester), E. E. Walker (Manchester), A. Wasserman (London), E. Heyman (London), P. Lewis Dale (London), A. M. Taylor (St Andrews), N. J. L. Megson (Teddington), L. C. Vernan (Teddington), W. P. Pepper (Liverpool), M. W. Perrin (Northwich), D. Findlayson (Long Eaton), J. Bouiton (Braintree), R. S. Morrel, W. R. Davies, J. Farquharson, H. I. Waterman, J. J. Leenderste, A. Abkin, S. Medvedev, H. Dorstal (Vienna), R. Freundenberg (Heidelberg), Dr & Mrs J. R. Katz (Amsterdam), K. H. Meyer (Geneva), G. Stafford Whitby (Ottawa), W. Herer (Freiburg), E. Husemann (Freiburg), H. B. Weiser (Houston, TX), W. O. Milligan (Houston, TX), J. C. Patrick (Yardville, NJ), G. Walter (Vienna), Dr & Mrs D. Signer (Bern), P. N. Kogerman (Tartu), E. Bergman (Rehevoth), D. L. Atwegg (Paris), T. F. Bradley (Linden, NJ), Dr & Mrs R. H. Kienle (Boundbrook, NJ), Fraulein D. D. Kruger (Berlin), M. Mathieu (Paris), R. Meyer (Paris), J. H. van der Monne (Amsterdam), Orthner (Frankfurt), H. Pringsheim (Prague), E. Proskauer (Leipzig), Sauter (Freiburg), T. Takei (Tokyo), P. von Tafel (Bern), J. C. Vluger (Amsterdam), H. I. Waterman (Delft), A. Weidinger (Amsterdam), F. Weigert (Leipzig).

Fig. 3.12 Attendees at the Faraday discussion on Polymerization (1935)

References

1. Sutton L, Davies M (1996) The history of the Faraday society. The Royal Society of Chemistry, London
2. van't Hoff JH (1888) Philos Mag 26:81
3. Pfeffer W (1877) Osmotische untersuchungen. Engelmann, Leipzig
4. Findlay A (1913) Osmotic pressure. Longmans, Green and Co., London
5. Staudinger H (1932) Die hochmolekularen Organischen Verbindungen. Springer, Berlin
6. Gibbs JW (1876) The equilibrium of heterogeneous substances. Trans Connect Acad Sci 3:228
7. Meyer KH, Mark H (1930) Der Aufbau der hochpolymeren organischen Naturstoffe. Akademische Verlagsgesellschaft, Leipzig
8. Mark H (1932) Physik and Chemie der Zellulose. Springer, Berlin
9. Semenov NN (1935) Chemical kinetics and chain reactions. Clarendon Press, Oxford

Chapter 4
Musings on the Prehistory of Polymer Science

The formation of a viable research paradigm requires many factors to be aligned. There needs to be a coherent collection of observable phenomena that requires the paradigm in order to comprehend the facts. Since there are polymers everywhere in the natural world, what set of observations provoked a search for an appropriate paradigm? The discovery of rubber elasticity by the European scientific community spurred the search for a way to comprehend the strange properties of this pure natural material.

All the powers of the French Academy of Sciences were needed: chemical analysis, mechanical analysis, thermodynamics, alchemy, synthesis. But even Berthelot could not really provide much insight. The Laws of Rubber Elasticity were established by Joule, but even the wizards of British physics could not conjure up an explanation. Until the dawning of the molecular age, no real progress was likely on this subject.

But, what is a molecule? Dalton had the vision to see heteronuclear diatomic molecules. Berthelot only had the vision to see synthesized substances of unknown structure. Kekule could imagine molecules linked together by chemical bonds with multifunctional atoms. The general formula for the n-alkanes, C_nH_{2n+2}, follows from the tetrafunctionality of carbon. The carbons form the backbone of the chain molecule and the hydrogens fill out the functionality. Van't Hoff had the vision to see the conformation of the molecule in three dimensions. Staudinger only had the insight gained by staring at molecular structures on a sheet of paper. He thought they were straight and two dimensional. Kuhn, Guth and Mark went where no chemist had gone before: chain molecules can be highly flexible and the number of distinguishable conformations is enormous. The world of molecular conformational entropy had arrived. Until the general theory of molecules had advanced to the present level, no solid foundation for a molecular science existed. Colloids could remain vague particles, albeit with variable surface charge, but the extremely detailed nature of molecular structure required more powerful microscopic tools.

G. Patterson, *A Prehistory of Polymer Science*, SpringerBriefs in History of Chemistry, 43
DOI: 10.1007/978-3-642-21637-4_4, © The Author(s) 2012

Early progress in molecular science is exemplified by the work of Michael Faraday. He was able to carry out good elemental analyses and to make good measurements of Dumas bulb gas densities. He discovered benzene in coal tar, and butene as a "polymer" of ethylene. But high molecular weight polymers are not volatile and another method was needed to obtain good molecular weights. Nevertheless, he was able to "see" natural rubber as a polymer of caoutchoucine. The concept of homologous series is one of the most powerful paradigms in polymer science. Haworth had the synoptic vision to encompass all the polysaccharides. There were differences in chain length, and differences in chain connectivity, but they were all carbohydrates. One of the most interesting aspects of the study of polymers as a function of chain length is the observation of crossover regions where the properties change qualitatively in addition to changing quantitatively. Low molecular weight n-alkane crystals are soft and often have several crystal forms. Higher molecular weights exhibit a series of phase transitions as the ultimate melting point is approached where the chains gain mobility around the chain axis but the overall parallel arrangement of the extended chains is retained. At higher molecular weights, the individual chains start to fold back on the crystal and individual crystal lamellae no longer reflect the fully extended length of the chains. At even longer lengths, individual chains are found in more than one lamella, and "tie chains" produce a tough material. All these substances have the same empirical formula, first exhibited by Kekule. Many more surprises await the diligent explorer of polymer crystals.

Jean Perrin searched all his life for ways to "see" atoms and molecules. The chance discovery of X-rays provided a key tool to help image the structure of molecules. While X-rays can be emitted and absorbed by atoms, they can also be scattered. Even if the structure of the liquid is isotropic, there will still be characteristic distances determined by the molecular structure. The molecules will be oriented at random overall and the X-ray scattering pattern will consist of radially symmetric intensity with variable magnitude. Amorphous polymers display the characteristic amorphous halos associated with liquids. Until a satisfactory theory of the liquid state could be formulated, no real progress in the understanding of bulk polymers could be made. Even late in the 20th century, some European scientists were denying the existence of the liquid state for polymers! As soon as crystalline polymers were examined with X-rays, it became clear that it was possible to have both high molecular weight and extensive crystallinity. The story of the slow realization that the "repeat unit" of the crystal could consist of parts of chains illustrates that the application of standard techniques to new materials can often lead to further insight into the nature of the technique. X-ray diffraction from polymer crystals became a highly skilled specialty. The variety of polymer crystal forms is huge and many phase transitions can be observed between different crystal forms. Polymers were studied near the start of X-ray diffraction, and they are studied today.

Another phenomenon that has been important to the understanding of polymers is viscosity. The spontaneous increase in viscosity of pure styrene liquids motivated scientists to find out why the viscosity was changing at constant temperature.

Berthelot recognized this fact as evidence of polymerization, but what did that mean? He also noted the evolution of heat during the thickening process. He correctly concluded that polymerization was a chemical reaction. The density of the system increases as the reaction proceeds. Viscosity increases with density in most liquids. But the increase was far too large to be due to densification alone. Einstein proved that solution viscosity increases as the particle size increases and as the volume fraction of particles increases. It was observed that some polymer particles were enormous, even though they weighed only a small fraction of the closest packed mass. Staudinger focused on the "size" and suggested that the contour length of a straight rod could produce large dimensions with a small effective spherically averaged density (long rods rotate in solution). But, when accurate molecular weights were obtained, and chemical structural models were examined, the measured particle radii were far too small to be due to fully extended rods. Staudinger could synthesize polymers of a wide range of types and sizes, but he had no facility in either mathematics or physics. Real science must either be carried out by polymaths (like James Clerk Maxwell) or better, by multidisciplinary teams. Mark, Polanyi, Kuhn and Guth made a great team.

Some polymers are very dense in solution and globular proteins are a good example of such a molecule. But for a "dense" system, it is less obvious that the structure is molecular, rather than merely colloidal. One of the key phenomena observed in protein solutions is denaturation. The solution viscosity undergoes a huge increase when the protein molecules unfold, due to heat or changes in salt or pH. If the protein globules were merely dense association complexes, the viscosity should be independent of the degree of association, since only the volume fraction of the solute matters. The observed increase in intrinsic viscosity is due to a true unfolding of the protein chains, covalently bonded together, even when "denatured." Even though the leaders in the protein community knew these facts, there were many members of that group that were not "molecular" in their thinking. Only when the protein scientists became thoroughly molecular did any progress occur.

The existence of a polymer industry provided a critical mass of workers who cared about the understanding of macromolecules. Both Mark and Carothers carried out seminal work in an industrial setting. Houwink used the Phillips Laboratory as a base of support. Baekeland was well trained as an academic chemist, and his desire to understand Bakelite included support for academic research. Polymer science as a distinct research community has always had a strong industrial component. Real materials provide both practical problems to be solved and new phenomena that need interpretation. The motivation to understand the items of technology led to the provision of resources to carry out investigations. During the 1930s, the number of trained academic chemists working in the polymer industry grew substantially.

In the most fundamental circles, macromolecules provided some of the most intriguing problems for deep analysis. The phenomenon of rubber elasticity is still inspiring theoretical work. The phenomenon of gelation and swelling is another classical problem for theorists. The multitude of crystal phases still excites crystallographers. Polymers also are involved in a wide range of ordered systems that are not locally crystalline, but are organized in regular lattices. When a field is understood at a level deep enough to attract the best theorists, a fruitful period can be expected. Werner Kuhn, Eugene Guth and Paul Flory were all captivated by the challenges of polymers.

All of human culture is captive to the flow of human history. The creation of a freely working community requires both a critical mass of workers and the freedom to communicate with one another on a regular basis. While there were fledgling polymer industries before the Great War, there were not enough people trained to think about materials in a molecular way to provide a critical mass. With a limited supply of actual scientists, competition for workers is intense. Physical chemistry was dominated by the Ionists: Wilhelm Ostwald, Jacobus van't Hoff and Svante Arrhenius. They had an entire field to found and chose solutions and chemical kinetics as the areas to pursue. Great physical chemists like Gilbert Lewis were occupied with thinking about small molecules; it is impossible to run, when it is already hard to walk! But, more general subjects like Thermodynamics were developed in such a way by Lewis and by Nernst, that when the time came to apply such analysis to polymers, Flory could simply rigorously proceed from first principles to the correct application. Einstein established the general principles of hydrodynamics, fluid fluctuations and heat capacity. When the time came to apply these classical theories to polymers, Kirkwood could simply proceed on a firm basis. Sometimes, an apparently quiet period is merely the gestation time for the foundation of a field.

After the War, there were many fewer scientists to engage in any activity. However, exciting new areas of spectroscopy, quantum mechanics and relativity were drawing some of the brightest and best workers. Nuclear science was exploding; the science of life was drawing Nobelists as fast as they could be minted. Natural selection applies to science itself! But, eventually, both the quantity and quality of scientists attracted to problems involving macromolecules increased until the human chain reaction took off. The discussions at the 1935 Discussion frequently led to "Aha!" moments, where auditors connected the dots during a lecture and announced exciting new conclusions. This positive feedback process further strengthened the community in its conviction that polymers were worth the effort.

In the natural history of life on Earth, many artifacts ended with only a brief lifespan. Many "fields" of science disappeared soon after they were announced by some prophet. Staudinger could have been one of those prophets, but in fact he persevered until his own mistaken thoughts were unable to derail the birth of polymer science. If colloid science had followed Wolfgang Ostwald, instead of The Svedberg, it would still be a swamp of irreproducible artifacts and useless speculations, expressed in a language that no outsider could comprehend. In fact, colloid science is today a thriving scientific community. Polymer science has become an even bigger community. But the future will be characterized by a multidisciplinary effort that includes all the fields noted in the Introduction.

Polymer science provides an interesting milieu in which to consider the nature of science and scientific communities. There is a unity to science that can be of great help to particular communities, if it is acknowledged and embraced. Paul Flory always stressed the unity of physical chemistry: it applied to colloidal systems, emulsions, polymers, gases, liquids, crystals, liquid crystals, plastic crystals and amorphous solids. Any attempt to "found" a new subfield of science that is disconnected from the rest of physical reality is doomed to failure. The community of polymer scientists chose to follow Carothers, Katz, Polanyi, Mark, Kuhn, Guth and Flory. These scientists were committed to the full molecular paradigm that applied to all molecules, not just to macromolecules. Since polymer science was strongly related to so many other subcommunities, any insight obtained in one community of discourse could be considered with respect to polymers, and vice versa. The Faraday Society was composed of all the leading physical chemists in England. They were committed to the application of physical chemistry to systems of real industrial and natural significance. Only by combining the insights of the best people and applying them to actual problems could genuine insight be obtained. When enough good people were discussing a particular problem, it was very common for new insights to be generated in real time. The commitment to include many good voices in the choir resulted in a scientific oratorio. This is still the case.

A progressive research community requires an actual working paradigm, but it is often instantiated in the minds of its leaders rather than explicitly written down. Michael Polanyi stressed the importance of tacit knowledge. This means that really good leaders are necessary. As long as polymer science consisted of a few fighting propagandists, no lasting progress could be made. Fortunately, leaders did emerge. They too had personal histories that included moments of epiphany when their commitment to the emerging paradigm was strengthened. Once the way forward is illuminated, many good people can run towards the light.

However, a few visionaries are not enough to firmly establish a field. There is an enormous amount of hard work needed to experimentally establish the actual behavior of nature. Until enough workers are available to carry out the work, any field will just be a glimmer in the dark. Polymer science had the advantage that, once the promise had been established, there were both governments and industries eager to spend money to develop applications of the newly established principles. The present story started with a consideration of caoutchouc. The story considered in this treatise also ends with the promise of synthetic rubber providing both money and motivation to invest in polymer science.

Biographical Sketch

Gary Patterson is Professor of Chemical Physics and Polymer Science at Carnegie Mellon University in Pittsburgh, Pennsylvania. He is also associated with the Chemical Heritage Foundation in Philadelphia as the Chief Bibliophile of the Bolton Society. He was educated at Harvey Mudd College (B.S. Chemistry, 1968) and Stanford University (Ph.D. Physical Chemistry, 1972). His thesis advisor was Paul Flory, who received the Nobel Prize for his work in Polymer Science in 1974. He was a Member of Technical Staff at the AT&T Bell Laboratories in the Chemical Physics Department from 1972–1984, when he joined Carnegie Mellon. He was the Charles Price Fellow of the Chemical Heritage Foundation in 2004–2005. He is a Fellow of the Royal Society of Chemistry, Faraday Division and the American Physical Society, Polymer Physics Division. He received the National Academy of Sciences Award for Initiatives in Research in 1981 for his work on the structure and dynamics of amorphous polymers. Gary is an active member of the HIST division of the American Chemical Society. His historical interests focus on polymer science and physical chemistry. His biographical interests include Michael Faraday, Jean Perrin and Paul John Flory. A biography of Flory is forthcoming.

G. Patterson, *A Prehistory of Polymer Science*, SpringerBriefs in History of Chemistry, 49
DOI: 10.1007/978-3-642-21637-4, © The Author(s) 2012